U.S. Department of Justice
National Institute of Justice

Vehicle Tracking Devices

NIJ Standard-0223.00

May 1986

ABOUT THE TECHNOLOGY ASSESSMENT PROGRAM

The Technology Assessment Program is sponsored by the Office of Development, Testing, and Dissemination of the National Institute of Justice (NIJ), U.S. Department of Justice. The program responds to the mandate of the Justice System Improvement Act of 1979, which created NIJ and directed it to encourage research and development to improve the criminal justice system and to disseminate the results to Federal, State, and local agencies.

The Technology Assessment Program is an applied research effort that determines the technological needs of justice system agencies, sets minimum performance standards for specific devices, tests commercially available equipment against those standards, and disseminates the standards and the test results to criminal justice agencies nationwide and internationally.

The program operates through:

The *Technology Assessment Program Advisory Council* (TAPAC) consisting of nationally recognized criminal justice practitioners from Federal, State, and local agencies, which assesses technological needs and sets priorities for research programs and items to be evaluated and tested.

The *Law Enforcement Standards Laboratory* (LESL) at the National Bureau of Standards, which develops voluntary national performance standards for compliance testing to ensure that individual items of equipment are suitable for use by criminal justice agencies. The standards are based upon laboratory testing and evaluation of representative samples of each item of equipment to determine the key attributes, develop test methods, and establish minimum performance requirements for each essential attribute. In addition to the highly technical standards, LESL also produces user guides that explain in nontechnical terms the capabilities of available equipment.

The *Technology Assessment Program Testing and Information Center* (TAPTIC) operated by a grantee, which supervises a national compliance testing program conducted by independent agencies. The standards developed by LESL serve as performance benchmarks against which commercial equipment is measured. The facilities, personnel, and testing capabilities of the independent laboratories are evaluated by LESL prior to testing each item of equipment, and LESL helps the Information Center staff review and analyze data. Test results are published in Consumer Product Reports designed to help justice system procurement officials make informed purchasing decisions.

Publications issued by the National Institute of Justice, including those of the Technology Assessment Program, are available from the National Criminal Justice Reference Service (NCJRS), which serves as a central information and reference source for the Nation's criminal justice community. For further information, or to register with NCJRS, write to the National Institute of Justice, National Criminal Justice Reference Service, Washington, DC 20531.

James K. Stewart, Director
National Institute of Justice

U.S. DEPARTMENT OF JUSTICE
National Institute of Justice

James K. Stewart, Director

ACKNOWLEDGMENTS

This standard was formulated by the Law Enforcement Standards Laboratory (LESL) of the National Bureau of Standards under the direction of Lawrence K. Eliason, Chief of LESL, and Marshall J. Treado, Program Manager for Communication Systems. P. Michael Fulcomer of the NBS Electrosystems Division was responsible for the preparation of this standard, with assistance from John R. Sorrells. The preparation of this standard was sponsored by the National Institute of Justice, Lester D. Shubin, Standards Program Manager. The standard has been reviewed and approved by the Technology Assessment Program Advisory Council.

The technical effort to develop this standard was conducted under Interagency Agreement LEAA-J-IAA-021-3, Project No. 8003.

FOREWORD

This document, NIJ Standard-0223.00, Vehicle Tracking Devices, is an equipment standard developed by the Law Enforcement Standards Laboratory of the National Bureau of Standards. It is produced as part of the Technology Assessment Program of the National Institute of Justice. A brief description of the program appears on the inside front cover.

This standard is a technical document that specifies performance and other requirements equipment should meet to satisfy the needs of criminal justice agencies for high quality service. Purchasers can use the test methods described in this standard to determine whether a particular piece of equipment meets the essential requirements, or they may have the tests conducted on their behalf by a qualified testing laboratory. Procurement officials may also refer to this standard in their purchasing documents and require that equipment offered for purchase meet the requirements. Compliance with the requirements of the standard may be attested to by an independent laboratory or guaranteed by the vendor.

Because this NIJ standard is designed as a procurement aid, it is necessarily highly technical. For those who seek general guidance concerning the selection and application of law enforcement equipment, user guides have also been published. The guides explain in nontechnical language how to select equipment capable of the performance required by an agency.

NIJ standards are subjected to continuing review. Technical comments and recommended revisions are welcome. Please send suggestions to the Program Manager for Standards, National Institute of Justice, U.S. Department of Justice, Washington, DC 20531.

Before citing this or any other NIJ standard in a contract document, users should verify that the most recent edition of the standard is used. Write to: Chief, Law Enforcement Standards Laboratory, National Bureau of Standards, Gaithersburg, MD 20899.

Lester D. Shubin
Program Manager for Standards
National Institute of Justice

NIJ STANDARD
FOR
VEHICLE TRACKING DEVICES

CONTENTS

COMMONLY USED SYMBOLS AND ABBREVIATIONS

A	ampere	H	henry	nm	nanometer
ac	alternating current	h	hour	No.	number
AM	amplitude modulation	hf	high frequency	o.d.	outside diameter
cd	candela	Hz	hertz (c/s)	Ω	ohm
cm	centimeter	i.d.	inside diameter	p.	page
CP	chemically pure	in	inch	Pa	pascal
c/s	cycle per second	ir	infrared	pe	probable error
d	day	J	joule	pp.	pages
dB	decibel	L	lambert	ppm	part per million
dc	direct current	L	liter	qt	quart
°C	degree Celsius	lb	pound	rad	radian
°F	degree Fahrenheit	lbf	pound-force	rf	radio frequency
diam	diameter	lbf·in	pound-force inch	rh	relative humidity
emf	electromotive force	lm	lumen	s	second
eq	equation	ln	logarithm (natural)	SD	standard deviation
F	farad	log	logarithm (common)	sec.	section
fc	footcandle	M	molar	SWR	standing wave ratio
fig.	figure	m	meter	uhf	ultrahigh frequency
FM	frequency modulation	min	minute	uv	ultraviolet
ft	foot	mm	millimeter	V	volt
ft/s	foot per second	mph	mile per hour	vhf	very high frequency
g	acceleration	m/s	meter per second	W	watt
g	gram	N	newton	λ	wavelength
gr	grain	N·m	newton meter	wt	weight

area = unit^2 (e.g., ft^2, in^2, etc.); volume = unit^3 (e.g., ft^3, m^3, etc.)

PREFIXES

d	deci (10^{-1})	da	deka (10)
c	centi (10^{-2})	h	hecto (10^2)
m	milli (10^{-3})	k	kilo (10^3)
μ	micro (10^{-6})	M	mega (10^6)
n	nano (10^{-9})	G	giga (10^9)
p	pico (10^{-12})	T	tera (10^{12})

COMMON CONVERSIONS
(See ASTM E380)

ft/s × 0.3048000 = m/s	lb × 0.4535924 = kg
ft × 0.3048 = m	lbf × 4.448222 = N
ft·lbf × 1.355818 = J	lbf/ft × 14.59390 = N/m
gr × 0.06479891 = g	lbf·in × 0.1129848 = N·m
in × 2.54 = cm	lbf/in^2 × 6894.757 = Pa
kWh × 3 600 000 = J	mph × 1.609344 = km/h
	qt × 0.9463529 = L

Temperature: $(T_F - 32) \times 5/9 = T_C$

Temperature: $(T_C \times 9/5) + 32 = T_F$

NIJ STANDARD
FOR
VEHICLE TRACKING DEVICES

1. PURPOSE AND SCOPE

The purpose of this document is to establish performance requirements and methods of test for vehicle tracking devices and systems. A system includes a receiver, a receiving antenna, a transmitter, a transmitting antenna, and a transmitter power source. Except for the receiving antenna, all parts of a vehicle tracking system are addressed by this standard. This standard applies to vehicle tracking devices or systems which either do not have special features such as audio alert tones or variable bandwidth receivers, or in which such special features are bypassed or disabled during testing for compliance with this standard. No type III vehicle tracking devices or systems were available at the time of development of this standard.

2. CLASSIFICATION

For the purpose of this standard, vehicle tracking devices or systems are classified by their operating frequencies.

2.1 Type I

Transmitters and receivers which operate in the 30–50 MHz band with a receiver channel spacing of 20 kHz.

2.2 Type II

Transmitters and receivers which operate in the 150–174 MHz band with a receiver channel spacing of 30 kHz.

2.3 Type III

Transmitters and receivers which operate in the 450–512 MHz band with a receiver channel spacing of 25 kHz.

3. DEFINITIONS

The principal terms used in this document are defined in this section. Additional definitions relating to law enforcement communications are given in LESP-RPT-0203.00, Technical Terms and Definitions Used with Law Enforcement Communications Equipment [1][1].

3.1 Adjacent-Channel Attenuation

A measure of the reduction in receiver response to an input signal whose frequency corresponds to a channel adjacent to that to which the receiver is tuned. It is expressed as the ratio of receiver response at the tuned frequency to the receiver response at that adjacent-channel frequency which produces the larger indication.

---

[1] Numbers in brackets refer to references in appendix A.

3.2 Average Carrier Output Power

For a pulsed-carrier transmitter, the power available at the output terminal averaged over the total operating time of the transmitter. It may be calculated as the product of peak carrier output power and duty cycle.

3.3 Carrier Frequency

The frequency generated by an unmodulated transmitter.

3.4 Directional Coupler

A transmission coupling device for separately sampling either the forward or backward wave in a transmission line. It is generally used to sample signals flowing in one direction while remaining isolated from those flowing in the opposite direction.

3.5 Duty Cycle

For a pulsed-carrier transmitter, the ratio of the time during which the maximum transmitter output signal is present at the output terminal to the total transmitter operating time. It may be calculated as the product of output pulse duration and the output pulse repetition rate.

3.6 Dynamic Range

For a receiver, the ratio between the maximum and the minimum input signals that will produce an output signal or indication within specified limits.

3.7 Figure of Merit

For this standard, a number obtained by combining transmitter peak output power and receiver direction indication sensitivity to form a vehicle tracking system performance rating.

3.8 Line Stretcher

A section of coaxial transmission line having an adjustable physical length.

3.9 Nominal Value

The numerical value of a device characteristic as specified by the manufacturer.

3.10 Peak Carrier Output Power

For a pulsed-carrier transmitter, the maximum instantaneous power of the periodically recurring transmitter output, as measured from its zero value and excluding relatively short duration switching transients.

3.11 Polarization

For an antenna, the orientation (horizontal, vertical, etc.) of the antenna elements to achieve a desired transmission or reception pattern.

3.12 Pulsed Carrier

Periodically recurring segments or pulses of a single unmodulated frequency generated by a transmitter. There is either no output or relatively little output from the transmitter between pulses.

3.13 Radiated Spurious Emission

Any emission from a transmitter at a frequency or frequencies which is not the specified operating frequency.

3.14 Radiation Efficiency

For an antenna, the ratio of the radiated power to the power delivered to the antenna, at a given frequency.

3.15 Sampler

A passive series device which couples energy over a broad frequency range from a transmission line into a third port. The attenuated output signal from the third port has the same waveform as the original signal.

3.16 Selectivity

The ability of a receiver to separate a desired signal frequency from other signal frequencies, some of which may differ only slightly from the desired signal.

3.17 Sensitivity

The minimum input signal required to produce a specified output signal or indication from a receiver.

3.18 Sideband Spectrum

For a transmitter, the frequency band, located immediately above and below the carrier frequency, which contains the emissions generated when the carrier frequency is modulated.

3.19 Signal Splitter

A passive series device which accepts an input signal and delivers multiple output signals which are isolated from one another and are of equal amplitude, phase, and waveform.

3.20 Spurious Response Attenuation

A measure of the reduction in receiver response to an input signal at a frequency which is both different from that to which the receiver is tuned and different from harmonics of the tuned frequency. It is expressed as the ratio of receiver response at the tuned frequency to the largest receiver response produced by any other frequency, except those that are harmonics of the tuned frequency.

3.21 Standard Short Circuit

A one-port load that terminates a section of coaxial transmission line in a short circuit.

3.22 Standing Wave Ratio (SWR)

The ratio of the maximum to the minimum amplitudes of the voltage or current appearing along a transmission line.

4. REQUIREMENTS

4.1 Minimum Performance

The vehicle tracking device performance shall meet or exceed the requirement for each characteristic as given below and in tables 1 through 5. These performance requirements meet or exceed those given in the Rules and Regulations published by the Federal Communications Commission (FCC) [2].

4.2 User Information

A nominal value for each of the characteristics listed in tables 1 through 5 shall be included in the information supplied to the purchaser by the manufacturer or distributor. In addition, the distributor shall provide:
   a) The range of temperatures within which the transmitter and receiver are designed to be operated.
   b) Transmitter and receiver operating frequencies.
   c) Peak and average rf carrier output power.

d) Current drain both during and between transmitted pulses.
e) Pulsed-carrier repetition rate and either pulse width or duty cycle.
f) Transmitter rf output impedance.
g) Receiver rf input impedance.
h) Receiver current drain and supply voltage.
i) Transmitter battery type and voltage.

4.3 Transmitter No-Load Characteristic

With the transmitter antenna disconnected, the transmitter shall operate for at least 1 min without any performance degradation. This characteristic, which shall be tested first, shall be measured in accordance with section 5.3.

4.4 Transmitter Load Condition

When tested under different load conditions in accordance with section 5.4, the output of the transmitter shall remain unconditionally stable and not change from a pulsed-carrier to a continuous wave.

4.5 Performance at Environmental Extremes

The ability of the vehicle tracking system to operate in environmental extremes shall be determined using the test methods described in section 5.5.

4.5.1 Temperature Stability

Low temperature tests of the transmitter and transmitter battery shall be conducted at -30 °C (-22 °F) or the lowest temperature at which the manufacturer states (sec. 4.2) that the equipment will operate properly, whichever is lower, and high temperature tests of the transmitter and transmitter battery shall be conducted at 60 °C (140 °F) or the highest temperature at which the manufacturer states that the equipment will operate properly, whichever is higher. The battery shall also be tested at 0 °C (32 °F).

Low temperature tests of the receiver shall be conducted at -10 °C (14 °F) or the lowest temperature at which the manufacturer states (sec. 4.2) that the equipment will operate properly, whichever is lower, and high temperature tests of the receiver shall be conducted at 50 °C (122 °F) or the highest temperature at which the manufacturer states that the equipment will operate properly, whichever is higher.

When the transmitter, transmitter battery, and receiver are tested in accordance with section 5.5.1 at the low and high temperatures defined above, their performance shall not fall outside the limits specified in items AK through AO, table 1, for transmitters; items CA, CC, and CD, table 3, for batteries; and items DG through DJ, table 4, for receivers. In addition, the transmitter/receiver figure of merit shall be at least 130 decibels above one milliwatt (dBm) (item EB, table 5).

TABLE 1. *Minimum performance requirements for transmitters used in vehicle tracking systems*[a]

Transmitter characteristic	Frequency band (MHz)	
	30–50	150–174
Radio Frequency Carrier Characteristics		
AA. Peak Carrier Output Power Variance	±1 dB	±1 dB
AB. Peak Output Power Variance (supply voltage varied +10% and −20%)	+1.5, −3 dB	+1.5, −3 dB
AC. Maximum Peak Carrier Output Power	1 W	1 W
AD. Maximum Average Carrier Output Power	30 mW	30 mW
AE. Carrier Frequency Variance (supply voltage varied +10% and −20%)	±0.002%	±0.002%
AF. Pulsed-Carrier Repetition Rate Variance (supply voltage varied +10% and −20%)	±5%	±5%
AG. Pulsed-Carrier Pulse Width Variance (supply voltage varied +10% and −20%)	±5%	±5%
Electromagnetic Compatibility Characteristics		
AH. Radiated Spurious Emission Attenuation	43 dB	43 dB
AI. Sideband Spectrum Attenuation (±1 kHz from carrier frequency)	25 dB	30 dB
AJ. Sideband Spectrum Attenuation (±2 kHz from carrier frequency)	50 dB	60 dB

Transmitter characteristic	Frequency band (MHz)	
	30–50	150–174
Temperature Stability: −30 and 60 °C (−22 and 140 °F)		
AK. Peak Output Power Variance	±3 dB	±3 dB
AL. Peak Output Power Variance (supply voltage varied +10% and −20%)	+3, −6 dB	+3, −6 dB
AM. Carrier Frequency Variance (supply voltage varied +10% and −20%)	±0.002%	±0.002%
AN. Pulsed-Carrier Repetition Rate Variance (supply voltage varied +10% and −20%)	±10%	±10%
AO. Pulsed-Carrier Pulse Width Variance (supply voltage varied +10% and −20%)	±10%	±10%
Humidity Stability: 90% rh at 50 °C (122 °F)		
AP. Peak Output Power Variance	±3 dB	±3 dB
AQ. Peak Output Power Variance (supply voltage varied +10% and −20%)	+3, −6 dB	+3, −6 dB
AR. Carrier Frequency Variance (supply voltage varied +10% and −20%)	±0.002%	±0.002%
AS. Pulsed-Carrier Repetition Rate Variance (supply voltage varied +10% and −20%)	±10%	±10%
AT. Pulsed-Carrier Pulse Width Variance (supply voltage varied +10% and −20%)	±10%	±10%
Vibration Stability		
AU. Peak Output Power Variance	±1 dB	±1 dB
AV. Carrier Frequency Variance	±0.002%	±0.002%
Dislodgement Characteristic		
AW. Force Required to Dislodge Transmitter	9 kg (19.8 lb)	9 kg (19.8 lb)

[a] Temperature at 25±5 °C (77±9 °F) and supply voltage within 1 percent of the nominal value unless otherwise indicated.

TABLE 2. *Minimum performance requirements for transmitter antennas used in vehicle tracking systems*[a]

Antenna characteristic	Frequency band (MHz)	
	30–50	150–174
BA. Radiation Efficiency	——	10%

[a] Temperature at 25±5 °C (77±9 °F) and supply voltage within 10 percent of specified nominal value.

TABLE 3. *Minimum performance requirements for transmitter batteries used in vehicle tracking systems*

Battery characteristic	Battery type	
	Mercury	Alkaline
Minimum Service Life		
CA. At 60±2 °C (140±3.6 °F)	60 h	36 h
CB. At 25±5 °C (77±9 °F)	60 h	30 h
CC. At 0±2 °C (32±3.6 °F)	12 h	15 h
CD. At −30±2 °C (−22±3.6 °F)	NA	3 h
Maximum Service Life		
CE. At all temperatures	10 d	10 d

4.5.2 Humidity Stability

The transmitter shall be tested at a temperature of 50 °C and at least 90 percent relative humidity and the receiver shall be tested at 50 °C and at least 85 percent relative humidity.

When the transmitter and receiver are tested in accordance with section 5.5.2 at the temperature and humidity conditions defined above, their performance shall not fall outside the limits specified in items AP through AT, for transmitters, and items DK through DN, for receivers. The transmitter/receiver figure of merit shall be at least 130 dBm (item EC).

4.5.3 Vibration Stability

When the transmitter and receiver are tested in accordance with section 5.5.3, no fixed part shall come loose, nor movable part be shifted in position or adjustment, as a result of the vibration. During the test, the peak output power and the carrier frequency shall not vary more than ±1 dB (item AU) and 0.002 percent (item AV) from the nominal value, respectively. Also, the direction indicator sensitivity shall not vary more than ±1 dB (item DO) from its nominal value and the adjacent-channel and spurious response attenuations shall be at least 60 dB (item DP) and 50 dB (item DQ), respectively.

4.5.4 Shock Stability

When the transmitter is tested in accordance with section 5.5.4, it shall suffer no more than superficial damage and no fixed part shall come loose, nor movable part be shifted in position or adjustment, as a result of the shock.

4.6 Transmitter Performance

4.6.1 Radio Frequency (RF) Carrier Characteristics

The rf carrier characteristics of output power, frequency stability, and pulsed-carrier repetition rate stability and pulse width stability shall be measured in accordance with section 5.6.1.

4.6.1.1 Output Power

When measured in accordance with section 5.6.1.1, the peak carrier output power delivered to a standard rf output load shall be within ±1 dB (item AA) of its nominal value when the transmitter supply voltage is within ±1 percent of its nominal value. When the transmitter supply voltage is increased by 10 percent or reduced by 20 percent of the nominal value, the peak carrier output power shall not vary more than +1.5 or − 3 dB (item AB). The maximum peak carrier output power and maximum average carrier output power shall not exceed 1 W (item AC) and 30 mW (item AD), respectively, as specified by the FCC [2].

Average carrier output power is calculated by multiplying peak carrier output power by the transmitter duty cycle. Transmitter duty cycle shall be determined in accordance with section 5.6.1.3.

4.6.1.2 Frequency Stability

When measured in accordance with section 5.6.1.2, and with the transmitter supply voltage varied from + 10 percent to −20 percent of the nominal value, the transmitter output frequency shall not vary from its nominal value by more than ±0.002 percent (item AE) as specified by the FCC [2].

4.6.1.3 Pulsed-Carrier Repetition Rate Stability and Pulse Width Stability

When measured in accordance with section 5.6.1.3, and with the transmitter supply voltage varied from + 10 percent to −20 percent of the nominal value, the transmitter pulsed-carrier repetition rate and pulse width shall not vary more than ±5 percent from nominal voltage (items AF and AG).

4.6.2 Electromagnetic Compatibility Characteristics

The electromagnetic compatibility characteristics of radiated spurious emissions and sideband spectrum shall be measured in accordance with section 5.6.2.

4.6.2.1 Radiated Spurious Emissions

When measured in accordance with section 5.6.2.1, each radiated spurious emission shall be attenuated a minimum of $43 + 10 \log_{10}$ (output power in watts) dB below the field strength of the carrier (item AH). This requirement fulfills certain FCC type-acceptance criteria [3].

4.6.2.2 Sideband Spectrum

When measured in accordance with section 5.6.2.2, each spurious sideband emission that is separated from the carrier frequency by ± 1 kHz or more shall be attenuated at least the amount indicated by item AI, table 1, and if separated from the carrier by ± 2 kHz or more, shall be attenuated at least the amount indicated by item AJ. These measurements comply with the FCC stipulation that the carrier shall have no modulation applied to carry information, and that its occupied bandwidth (containing 99% of the radiated power) shall not exceed 2 kHz [2].

4.6.3 Dislodgement Characteristic

When measured in accordance with section 5.6.3, a force of at least 9 kg (19.8 lb) (item AW), averaged over three trials, shall be required to dislodge the vehicle tracking transmitter after it has been attached in the prescribed manner to a vertical unpainted steel plate.

4.7 Transmitter Antenna Radiation Efficiency (Type II Only)

When measured in accordance with section 5.7, the transmitter antenna radiation efficiency shall be at least 10 percent (item BA, table 2).

4.8 Transmitter Battery Service Life

The transmitter battery characteristics of minimum and maximum battery service life shall be measured in accordance with section 5.8. The transmitter shall operate for at least the number of hours indicated by item CB, before its peak output power decreases 3 dB. The transmitter shall operate no longer than 10 days before its peak power decreases 10 dB (item CE).

TABLE 4. *Minimum performance requirements for receivers used in vehicle tracking systems[a]*

Receiver characteristic	Frequency band (MHz)	
	30–50	150–174
Sensitivity Characteristics		
DA. Direction Indication Sensitivity	−113 dBm	−113 dBm
DB. Relative Distance Indication Sensitivity	−103 dBm	−103 dBm
Selectivity Characteristics		
DC. Adjacent-Channel Attenuation	60 dB	60 dB
DD. Spurious Response Attenuation	50 dB	50 dB
Dynamic Range Characteristic		
DE. Minimum Dynamic Range	90 dB	90 dB
Temperature Stability: −10 and 50 °C (14 and 122 °F)		
DG. Direction Indication Sensitivity Variance	+3, −6 dB	+3, −6 dB
DH. Adjacent-Channel Attenuation	60 dB	60 dB
DI. Spurious Response Attenuation	50 dB	50 dB
DJ. Minimum Dynamic Range	90 dB	90 dB
Humidity Stability: 85% rh at 50 °C (122 °F)		
DK. Direction Indication Sensitivity Variance	+3, −6 dB	+3, −6 dB
DL. Adjacent-Channel Attenuation	60 dB	60 dB
DM. Spurious Response Attenuation	50 dB	50 dB
DN. Minimum Dynamic Range	90 dB	90 dB
Vibration Stability		
DO. Direction Indication Sensitivity Variance	±1 dB	±1 dB
DP. Adjacent-Channel Attenuation	60 dB	60 dB
DQ. Spurious Response Attenuation	50 dB	50 dB

[a] Temperature at 25±5 °C (77±9 °F) and supply voltage between 11.4 and 14.4 V dc unless otherwise indicated.

4.9 Receiver Performance

4.9.1 Sensitivity Characteristics

The sensitivity characteristics of direction indication and distance indication shall be measured in accordance with section 5.9.1.

4.9.1.1 Direction Indication Sensitivity

When the receiver is tested in accordance with sections 5.9.1 and 5.9.1.1 at standard supply voltages of 11.4 to 14.4 V dc, a receiver input signal of -113 dBm (item DA) shall produce a direction indication equal to at least 10 percent of full scale or twice the amplitude of random noise variations, whichever is greater.

4.9.1.2 Distance Indication Sensitivity

When the receiver is tested in accordance with sections 5.9.1 and 5.9.1.2 at standard supply voltages of 11.4 to 14.4 V dc, a receiver input signal of -103 dBm (item DB) shall produce a distance indication equal to at least 10 percent of full scale.

4.9.2 Selectivity Characteristics

The selectivity characteristics of adjacent-channel attenuation and spurious response attenuation shall be measured in accordance with section 5.9.2.

4.9.2.1 Adjacent-Channel Attenuation

When the receiver is tested in accordance with sections 5.9.2 and 5.9.2.1 at standard supply voltages of 11.4 to 14.4 V dc, the adjacent-channel attenuation shall be at least 60 dB (item DC).

4.9.2.2 Spurious Response Attenuation

When the receiver is tested in accordance with sections 5.9.2 and 5.9.2.2 at standard supply voltages of 11.4 to 14.4 V dc, the spurious response attenuation shall be at least 50 dB (item DD).

4.9.3 Dynamic Range Characteristic

When the receiver is tested in accordance with section 5.9.3 at standard supply voltages of 11.4 to 14.4 V dc, the dynamic range shall be at least 90 dB (item DE) without overloading the distance indicating device.

TABLE 5. *Minimum system performance requirements for vehicle tracking systems*[a]

System characteristic	Frequency band (MHz)	
	30–50	150–174
Transmitter/Receiver Figure of Merit		
EA. Figure of Merit at 25±5 °C (77±9 °F)	140 dBm	140 dBm
EB. Figure of Merit with Transmitter at −30 °C (−22 °F) or 60 °C (140 °F), and Receiver at −10 °C (14 °F) or 50 °C (122 °F)	130 dBm	130 dBm
EC. Figure of Merit with Transmitter at 90% rh and 50 °C (122 °F), and Receiver at 85% rh and 50 °C (122 °F)	130 dBm	130 dBm

[a] Transmitter supply voltage at 20 percent below nominal value and receiver supply voltage between 11.4 and 14.4 V dc.

4.10 Transmitter/Receiver Figure of Merit

When calculated in accordance with section 5.10, the transmitter/receiver figure of merit shall be at least 140 dBm (item EA).

5. TEST METHODS

5.1 Standard Test Conditions

Allow all measurement equipment to warm up until it has achieved sufficient stability to perform the tests without adding errors due to calibration drift. Check the calibration of all equipment after warm-up and prior to testing. Recheck calibration at the conclusion of the measurements to verify that no drift has occurred that would affect the accuracy of the results. Unless otherwise specified, perform all measurements under the following standard test conditions.

5.1.1 Standard Temperature

Standard ambient temperature shall be between 20 and 30 °C (68 and 86 °F).

5.1.2 Standard Relative Humidity

Standard ambient relative humidity shall be between 20 and 70 percent.

5.1.3 Standard Supply Voltage

5.1.3.1 Transmitter Standard Supply Voltage

The standard supply voltage for transmitters shall be within ±1 percent of the nominal battery voltage (sec. 4.2). When variation in supply voltage is required, tests shall be performed using a well-filtered, variable, electronic regulated, dc supply. When battery operation is required, tests shall be performed using a fully-charged battery of the same type normally used in the transmitter.

5.1.3.2 Receiver Standard Supply Voltage

The standard supply voltage for receivers shall be within ±1 percent of the nominal supply voltage (sec. 4.2). Several tests require that the supply voltage be varied over a specified range and that data be taken at the supply voltage which produces the poorest results. A well-filtered, variable, electronic regulated, dc supply shall be used for such tests.

5.1.4 Standard Test Frequencies

The standard test frequencies shall be the transmitter and receiver operating frequencies (sec. 4.2).

5.1.5 Standard Radiation Test Site

The standard radiation test site shall be located on level ground which has uniform electrical characteristics (i.e., ground constants). Reflecting objects (especially large metal objects), trees, buildings, and other objects which would perturb the electromagnetic fields to be measured should not be located closer than 90 m (295 ft) from any test equipment or equipment under test. All utility lines and any control circuits between test positions should be buried underground. The ambient electrical noise level should be as low as possible and shall be carefully monitored to ensure that it does not interfere with the test being performed. Preferably, the test site should be equipped with a remotely-controlled turntable located at ground level.

5.2 Test Equipment

The test equipment discussed in this section is that equipment which is the most critical in making the measurements required by this standard. All other test equipment shall be of comparable quality. Each piece of test equipment shall have been calibrated with accuracy traceable to a recognized standard.

5.2.1 Spectrum Analyzer

The spectrum analyzer shall have a 50-Ω input impedance, a frequency range from at least 0.1 to 1000 MHz, and a variable persistence storage display. It shall provide a vertical logarithmic display with at least two selectable sensitivities, e.g., 2 dB and 10 dB per division. The display shall be capable of being calibrated to an accuracy of 1.5 dB maximum over the 10 dB per division range and 0.5 dB maximum over the 2 dB per division range. It shall accept an input of at least 30 dBm peak and be equipped with a calibrated 10 dB per step attenuator. It shall be able to resolve signals that are 300 Hz apart at 6 dB below the maximum signal amplitude and 1.2 kHz apart at 60 dB below the maximum signal amplitude. It shall provide a means of identifying a specific frequency on the display to an accuracy of at least ± 10 MHz with a resolution of at least 1 MHz.

5.2.2 Radio Frequency Power Meter

The rf power meter shall have an input impedance of 50-Ω, a maximum SWR of 1.20, and a frequency range which spans both the frequency of the transmitter to be tested and the frequency of any calibration source. It shall provide an internal power reference source with an accuracy traceable to the National Bureau of Standards of ± 1 percent for use in internal or external calibrations. After calibration, the meter shall have an accuracy of at least 2 percent of full scale. It shall have an input amplitude range which can accommodate the maximum output from the transmitter to be tested.

5.2.3 Radio Frequency Signal Generator

The rf signal generator must be capable of providing bursts or pulses of rf sine waves in addition to the normal continuous wave operation. The generator output amplitude between bursts shall be less than 1 percent of the burst amplitude. The peak output signal amplitude should not change by more than 2 percent when switching between burst operation and continuous wave operation.

The signal generator shall have an output impedance of 50-Ω and a frequency range from at least 0.1 to 500 MHz with 1000 MHz being more desirable. It shall have a calibrated variable output level accurate to ± 0.2 dB per 10 dB step when terminated in a 50-Ω load, and shall cover a range of at least -120 to 0 dBm. The output level shall remain flat to within ± 1.0 dB relative to the 50 MHz level when the frequency is increased to 500 MHz, and flat to within ± 2.5 dB relative to the 50 MHz level when increased from 500 to 1000 MHz. The amplitude of all harmonics should be at least 30 dB below that of the carrier frequency, and the amplitude of other spurious signals that are separated from the carrier by 20 kHz or more should be at least 70 dB below that of the carrier frequency.

The generator should include a digital frequency counter which has an uncertainty of no greater than one part in 10^6. If an integral frequency counter is not included, a separate unit having the required accuracy shall be provided. This counter is in addition to the one specified below.

5.2.4 Variable DC Power Supply

The variable dc power supply shall have at least 0.1 percent regulation when line or load changes.

5.2.5 Radio Frequency Counter

The rf counter shall be capable of measuring the frequency of a pulsed rf carrier to an uncertainty of no greater than one part in 10^6 for the narrowest transmitter pulse expected to be encountered.

5.2.6 Oscilloscope

The oscilloscope shall have a calibrated time base, a bandwidth of at least 10 MHz and a variable persistence storage display.

5.2.7 Coaxial Radio Frequency Detector

The coaxial rf detector shall have input and output impedances of 50-Ω and an input SWR of less than 1.4. It shall either (1) have a maximum input rating sufficiently high to accept the output of the transmitter under test directly, or (2) be preceded by coaxial attenuators to reduce the transmitter voltage to a value that will not damage the detector. The detector shall have a frequency range extending from below 200 kHz to above that of the highest transmitter frequency to be measured.

5.2.8 Field Strength Meter

The field strength meter, consisting of an antenna and a shielded, calibrated receiver, shall be capable of portable operation and shall cover the frequency range from at least 0.1 to 1000 MHz. It shall be able to measure a pulsed rf signal to an accuracy of ± 2 dB at the lowest transmitter duty cycle expected to be encountered, and shall have a resolution of at least 0.3 dB. It should be able to measure signals as small as one microvolt per meter, sometimes expressed as 0 dB above one microvolt per meter (dBμV/m).

NOTE 1: A peak detection circuit is normally used to enable measurement of the maximum value of a pulsed rf signal. Some instruments may use or offer the selection of a "quasi-peak" detection circuit, however. A "quasi-peak" detection circuit does not indicate the true peak amplitude for pulses with a repetition rate less than 1 kHz. A correction factor, which depends upon the pulse repetition rate (and which should be specified in the equipment instruction manual), must be added if "quasi-peak" detection is utilized.

5.2.9 Pulse Generator[2]

The pulse generator shall be capable of duplicating the pulse width and pulse repetition rate of the transmitter(s) to be tested, and be capable of modulating the rf signal generator (see sec. 5.2.3). The pulse generator itself need not be capable of reproducing the very low repetition rates at which some transmitters operate as long as it can be triggered at the required rate by another device, such as a function generator, having the required capability.

5.2.10 Radio Frequency Signal Splitter

The rf signal splitter, which is often part of a splitter/combiner, shall have a 50-Ω impedance at all terminals, an SWR of less than 1.3 and a frequency range extending from below 200 kHz to above that of the highest receiver frequency to be measured. Its amplitude imbalance between channels shall be no greater than 0.2 dB and the isolation between each output port shall be a minimum of 30 dB. The number of outputs required for the splitting function depends on the number of inputs required by the receiver under test (see sec. 5.9.1).

5.2.11 Environmental Chamber(s)

The environmental chamber or chambers shall be capable of providing air temperatures over a range of at least −30 to 60 °C (−22 to 140 °F) and relative humidities up to at least 90 percent at 50 °C (122 °F). During tests, the item under test shall be shielded from air currents flowing directly from heating or cooling elements in the chamber. The temperature of the test item shall be measured by an instrument that is separate from the sensor used to control the chamber air temperature. Likewise, humidity shall be measured with a hygrometer that is separate from the sensor used to control humidity.

5.2.12 Vibration Machine

The vibration machine shall have the capacity to accept the transmitter or receiver to be tested. It shall be able to produce simple harmonic motion having (1) a total excursion of 0.76 mm (0.03 in) at frequencies between 10 and 30 Hz, and (2) a total excursion of 0.38 mm (0.015 in) at frequencies between 30 and 60 Hz. A sine-wave generator with less than 1 percent distortion and, in many cases, a power amplifier will be necessary to drive the vibration machine.

5.2.13 Standard Radio Frequency Output Load (Transmitter)

The standard rf output load shall be a 50-Ω resistive termination having an SWR of 1.1 or less at the standard test frequencies. If connectors and cables are used to attach the standard output load to the transmitter, the combined SWR, including the load, shall be 1.1 or less.

5.3 Transmitter No-Load Test

Disconnect the transmitter antenna from the transmitter, and install the batteries in the transmitter. Switch on the transmitter and operate it for at least 1 min. Wait at least 20 min and then proceed with the next test.

---

[2] The purpose of this generator is to provide pulse modulation for the rf signal generator specified in section 5.2.3. If the rf signal generator can provide burst or pulse operation without external modulation or if other suitable methods are available, this generator is not necessary.

5.4 Transmitter Load Condition Test

Connect the transmitter to the test equipment as shown in figure 1, step 1. Turn on the transmitter and adjust the analog power meter until a maximum indication is observed at approximately midscale on the meter. The indication on the meter scale will not remain constant but rather will fluctuate back and forth between a maximum and minimum indication as long as the transmitter output remains in a pulsed-carrier condition.

Replace the standard output load with a standard short circuit as in step 2 and vary the line stretcher over one-half wavelength. At the same time observe the indication on the meter scale. If the indication on the scale does not continue to fluctuate, the transmitter fails the test. This indicates the output of the transmitter has. changed from a pulsed-carrier to a continous wave.

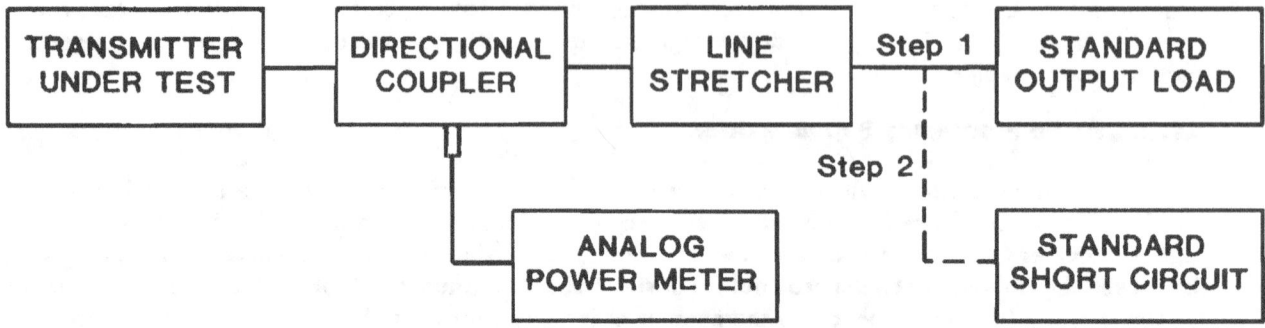

FIGURE 1. *Block diagram for the transmitter load condition measurement.*

5.5 Transmitter and Receiver Environmental Tests

5.5.1 Temperature Test

Using a variable dc power supply, place the transmitter, with power turned off, in the environmental chamber. Adjust the chamber to within ±2 °C (±3.6 °F) of the required low temperature. Allow the transmitter to reach temperature equilibrium and maintain it at this temperature for 30 min. With the transmitter still in this environment, turn on the power supply. Five minutes after turn-on, test the transmitter in accordance with the appropriate procedures of section 5.6. Return to ambient temperature and repeat the above procedure at the required high temperature ±2 °C (±3.6 °F).

Repeat this procedure for the receiver at its required low and high temperatures ±2 °C (±3.6 °F), testing in accordance with the appropriate procedures of section 5.9.

For transmitter battery tests, adjust the environmental chamber to within ±2 °C (±3.6 °F) of the required high temperature. With the transmitter switched off, install the battery or batteries to be tested in the transmitter and connect the spectrum analyzer to the transmitter output. After the environmental chamber has stabilized at the desired temperature, wait 30 min, switch on the transmitter and place it in the chamber. Test the transmitter batteries in accordance with section 5.8. Repeat this procedure for each of the other required temperatures.

5.5.2 Humidity Test

Using a variable dc power supply, place the transmitter, with power turned off, in the environmental chamber. Adjust the chamber to a relative humidity of 90 percent at a temperature of 50 °C (122 °F) and maintain the transmitter at these conditions for at least 8 h. With the transmitter still in this environment, turn on the power supply. Five minutes after turn-on, test the transmitter in accordance with the appropriate procedures of section 5.6.

Repeat this procedure for the receiver at 85 percent relative humidity and 50 °C (122 °F), testing in accordance with section 5.9.

5.5.3 Vibration Test

Power for the transmitter shall be furnished by batteries of the type normally used in the equipment. Receiver supply voltage may be furnished by a regulated dc power supply. Fasten the device to be tested to the vibration machine with a rigid mounting fixture.

Perform a two-part test by subjecting the transmitter to controlled vibrations in each of three mutually perpendicular directions, one of which is the vertical. Testing time for each direction is 30 min.

First subject the transmitter to three 5-min cycles of simple harmonic motion having an amplitude of 0.38 mm (0.015 in) [total excursion of 0.76 mm (0.03 in)] applied initially at a frequency of 10 Hz and increased at a uniform rate to 30 Hz in 2-1/2 min, then decreased at a uniform rate to 10 Hz in 2-1/2 min.

Then subject the unit to three 5-min cycles of simple harmonic motion having an amplitude of 0.19 mm (0.0075 in) [total excursion of 0.38 mm (0.015 in)] applied initially at a frequency of 30 Hz and increased at a uniform rate to 60 Hz in 2-1/2 min, then decreased at a uniform rate to 30 Hz in 2-1/2 min.

Repeat the above for each of the other two directions. Measure the performance of the transmitter during vibration, in accordance with the appropriate procedures of section 5.6.

Repeat this procedure for the receiver, testing in accordance with section 5.9.

5.5.4 Shock Test

·With its antenna disconnected, drop the transmitter once on each of four or more sides (all sides not having a protrusion or antenna connection), from a height of 1 m (3.28 ft) onto a smooth concrete floor. Turn off the transmitter power during the test; if necessary, use appropriate guides to ensure contact with the floor by the desired equipment surface.

5.6 Transmitter Tests

For all tests in which the transmitter output is connected via coaxial cable to a test instrument, the test instrument input impedance must equal the standard rf output load for the transmitter or a separate rf output load must be used with a sampler inserted between the transmitter and load to derive a signal for the test instrument. The test setup diagrams in this document assume that any test instrument to which the transmitter output is connected has a 50-Ω input impedance.

5.6.1 Radio Frequency Carrier Tests

5.6.1.1 Peak Output Power Test

Since most power meters will not respond correctly to the pulsed-carrier output that is characteristic of a vehicle tracking system transmitter, a spectrum analyzer is used to measure peak output power.

Before use, calibrate the spectrum analyzer using the test setup shown in step 1, figure 2. Set the rf signal generator for continuous wave operation and adjust the signal generator frequency to equal that of the transmitter under test. Set the signal generator output to two or three different amplitude levels in the same

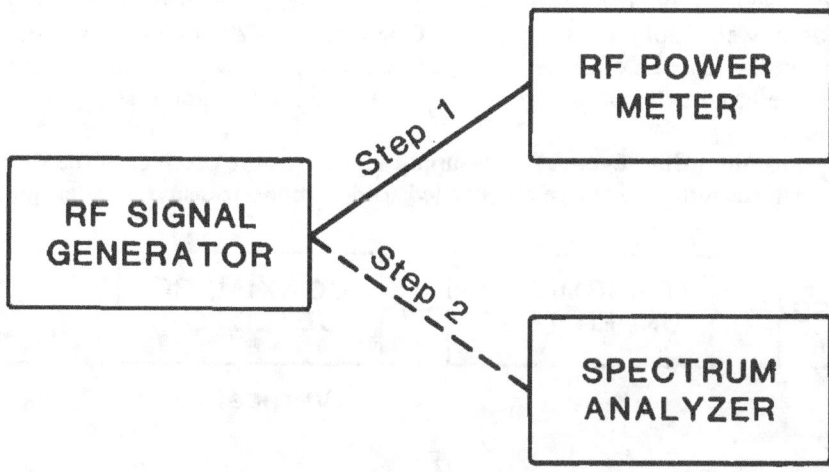

FIGURE 2. *Block diagram for calibration of the spectrum analyzer to measure peak power.*

general range as that expected from the transmitter, and measure the output at each level with the rf power meter. Then disconnect the signal generator from the rf power meter and connect it to the spectrum analyzer (step 2) and measure the same signal generator output levels. If necessary, either calculate a correction factor for the spectrum analyzer display or readjust the spectrum analyzer calibration. If a large discrepancy exists between the rf power meter and the spectrum analyzer, identify the source of error and correct it before proceeding.

Connect the transmitter and test equipment as shown in figure 3, adjust the variable dc power supply to the transmitter standard supply voltage, and measure the peak output power on the spectrum analyzer. Record this value for use in section 5.7. Then change the standard supply voltage +10 percent and −20 percent, and measure peak output power at each setting.

Calculate average output power as the product of peak output power and the transmitter duty cycle. (See sec. 5.6.1.3.)

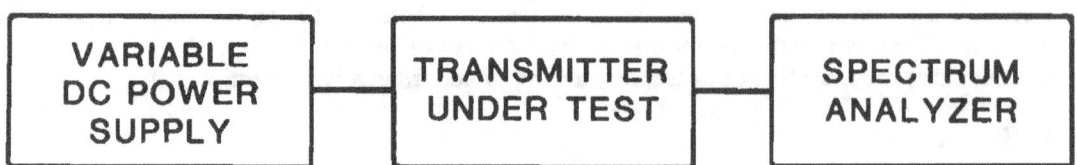

FIGURE 3. *Block diagram for peak output power and sideband spectrum measurements.*

5.6.1.2 Frequency Stability Test

Connect the transmitter and test equipment as shown in figure 4. Adjust the variable dc power supply to the transmitter standard supply voltage and measure the transmitter frequency. Change the standard supply voltage +10 percent and −20 percent and again measure the frequency. Note the largest deviations from the nominal transmitter frequency (sec. 4.2.b) and calculate the percent deviation.

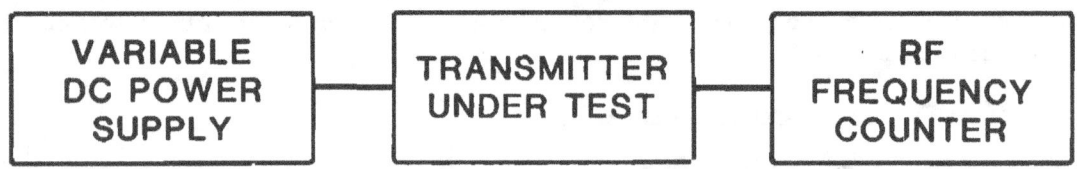

FIGURE 4. *Block diagram for frequency stability measurement.*

5.6.1.3 Pulsed-Carrier Repetition Rate Stability, Pulsed-Carrier Pulse Width Stability, and Transmitter Duty Cycle Tests

Connect the oscilloscope to the transmitter output via a coaxial rf detector as shown in figure 5. Only the envelope of the pulsed rf waveform, as provided by the coaxial rf detector, is necessary for these measurements.

Vary the dc power supply voltage from +10 percent to −20 percent of the transmitter standard supply voltage, making measurements of pulse repetition rate and pulse width at nominal voltage and at the voltage extremes. Note the largest deviations from the value measured at nominal supply voltage and calculate the percent deviation.

Calculate transmitter duty cycle at each supply voltage as the product of the pulse repetition rate and the pulse width. The transmitter duty cycle is needed to determine transmitter average output power in section 5.6.1.1.

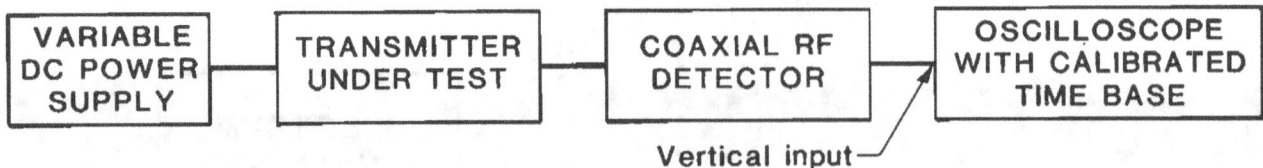

FIGURE 5. *Block diagram for pulsed-carrier repetition rate stability, pulsed-carrier pulse width stability, and transmitter duty cycle measurements.*

5.6.2 Electromagnetic Compatibility Tests

5.6.2.1 Radiated Spurious Emissions Test

If the field test site used for this test has 120 V dc power available, perform the measurements using the test setup shown in figure 6, utilizing a spectrum analyzer with variable persistence storage display instead of the field strength meter. Follow the procedures described in the second step of this test procedure, except that the frequency of each radiated spurious emission will not be known.

If the field test site used for this test does not have 120 V dc power available (battery-powered spectrum analyzers were not commercially available as of the date this standard was prepared) perform the measurements in two steps, the first in the laboratory and the second at the standard radiation test site (sec. 5.1.5), using a battery-powered field strength meter. First, make a preliminary measurement in the laboratory using the spectrum analyzer method to determine the approximate frequency of any high amplitude spurious signals radiated by the transmitter under test. Use this information to facilitate spurious signal measurements at the standard radiation test site.

For the preliminary measurement, set up the equipment as shown in figure 6. Attach the transmitter, operated from its battery and with its antenna connected, to a metal surface. For type II devices, connect a whip antenna approximately one-quarter wavelength long to the spectrum analyzer input. For type I devices, a shorter wavelength antenna may be used—possibly one of the vehicle tracking receiver antennas. The spectrum analyzer antenna may be mounted on the same surface as the transmitter, if desired, with coaxial cable connecting it to the analyzer input. The transmitter and spectrum analyzer antennas should be polarized in the same direction, located .9 to 1.2 m (3 to 4 ft) apart and as far as possible from other large metal objects. Set the spectrum analyzer display to show all frequencies from the lowest radio frequency generated by the transmitter to the 10th harmonic of the carrier or 1000 MHz, whichever is higher.

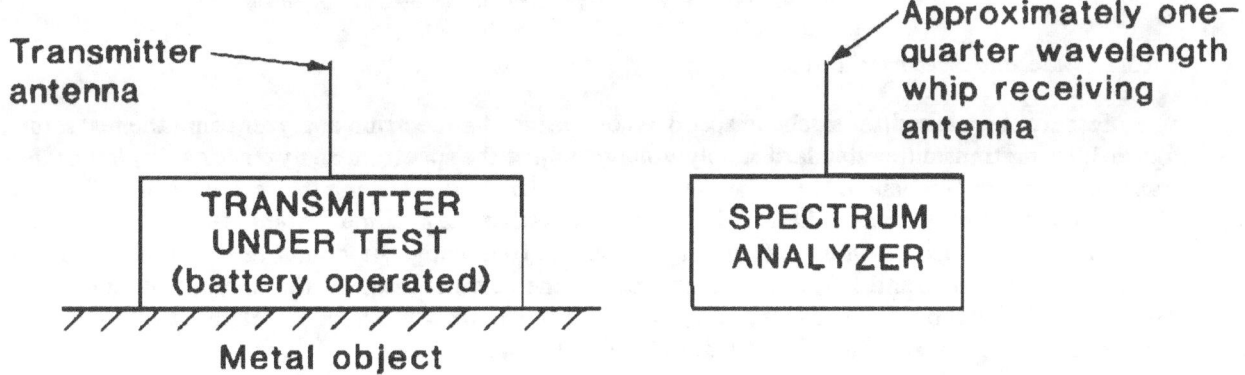

FIGURE 6. *Block diagram for preliminary radiated spurious emissions measurement.*

Turn on the transmitter and adjust the spectrum analyzer display to show the spurious signals. Next, turn off the transmitter and note any signals remaining on the display. Since these were not generated by the transmitter, they should be ignored when making final measurements. Reactivate the transmitter and record the frequency and relative amplitude of each transmitter-radiated spurious emission.

Measure the radiated spurious emissions using the information obtained above. Perform this second step at the standard radiation test site, setting up the equipment as shown in figure 7. Place the transmitter, with antenna connected, on a nonmetallic support so that the antenna is horizontally polarized and its tip is approximately 127 cm (50 in) above the earth. Place the halfwave dipole receiving antenna so that it is also horizontally polarized and located 30 m (98.4 ft) from the transmitter and 3.0 m (9.8 ft) above the earth. Turn on the transmitter and measure the field strength of the carrier frequency in dBμV/m using the technique described below. Record this value for use in section 5.7. Then, if available, use the frequency data obtained in step one as a guide to locate and measure the field strength in dBμV/m of any radiated spurious emissions, from the lowest radio frequency generated by the transmitter to at least 1000 MHz.

In the field, spurious emissions are most likely to be detected at those frequencies which produced the highest level spurious signals in the preliminary laboratory tests. If a tunable receiving antenna is used, adjust the antenna to a half wavelength for each frequency measured. If a series of fixed-broadband antennas is used, include the antenna correction factor when calculating spurious attenuation for each frequency.

For the carrier and each spurious frequency, reposition the horizontally-polarized receiving antenna a quarter wavelength in any direction to obtain a maximum reading on the field strength meter. Rotate the transmitter to further increase this maximum reading. Repeat this procedure of moving the receiving antenna (both back and forth and up and down) and rotating the transmitter until the largest signal has been obtained and recorded. Then place the receiving antenna in a vertical position and repeat the procedure for each spurious signal.

The attenuation of each radiated emission is the field strength in dBμV/m of the carrier frequency minus the field strength in dBμV/m of the radiated spurious emission.

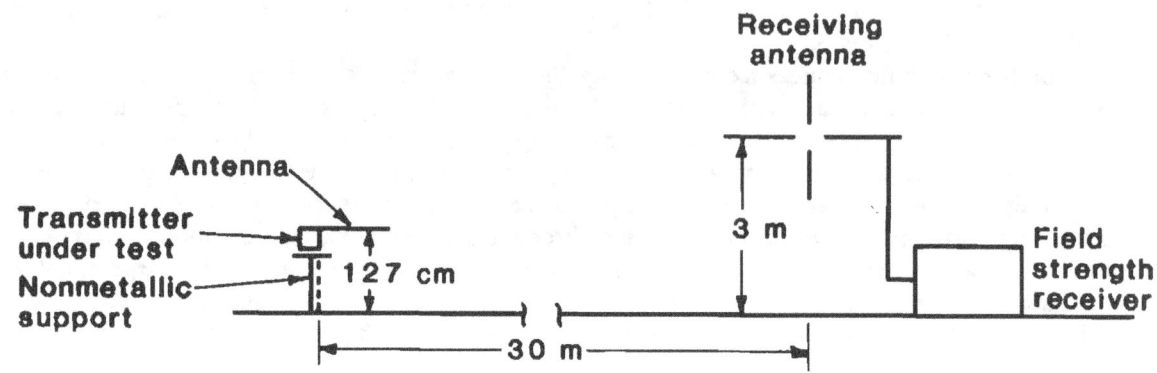

FIGURE 7. *Block diagram for radiated spurious emission measurement.*

5.6.2.2 Sideband Spectrum Test

Measure the transmitter sideband spectrum by means of a spectrum analyzer using the test setup shown in figure 3. Using transmitter standard supply voltage, adjust the spectrum analyzer for a display centered on the transmitter carrier frequency. Decrease the frequency span, while keeping the carrier centered, so that the total display covers no more than 10 kHz. Adjust the trace so that the maximum point just touches the top reference line (see fig. 8). This maximum occurs at the carrier frequency. At points on the trace corresponding to frequencies that are 1 and 2 kHz above and below the carrier, measure the amount of signal reduction, in decibels, from the top reference line (see fig. 8). Note the *minimum* reduction relative to the carrier amplitude for the ±1-kHz separation and for the ±2-kHz separation.

5.6.3 Dislodgement Test

Attach the vehicle tracking transmitter to a vertical unpainted, uncoated 0.3 to 1.0 cm (1/8 to 3/8 in) thick steel plate by the same method used to attach the transmitter to a vehicle. Locate the transmitter above a padded horizontal surface to cushion it if it falls.

Loop a cord with a working load limit of at least 22.7 kg (50 lb) around the transmitter in a manner that will enable the unit to be pulled outward in a direction perpendicular to the steel plate from a point approximately in line with the transmitter center of gravity. Apply the pull force gradually and measure it with a push-pull force gauge having a maximum reading pointer and a capacity of at least 14.0 kg (31 lb). Provide a vertical padded guard to prevent the transmitter from striking the force gauge or the person performing the test if it dislodges from the plate.

Perform three separate tests, measuring the pull force required to dislodge the transmitter from the steel plate for each trial. Reject the transmitter if it is dislodged by a pull of less than 4.5 kg (10 lb) for any trial. *Do not exceed a pull of 14.0 kg (31 lb).* Average the three results for the final determination, assigning a value of 14 kg to any trial exceeding 14 kg.

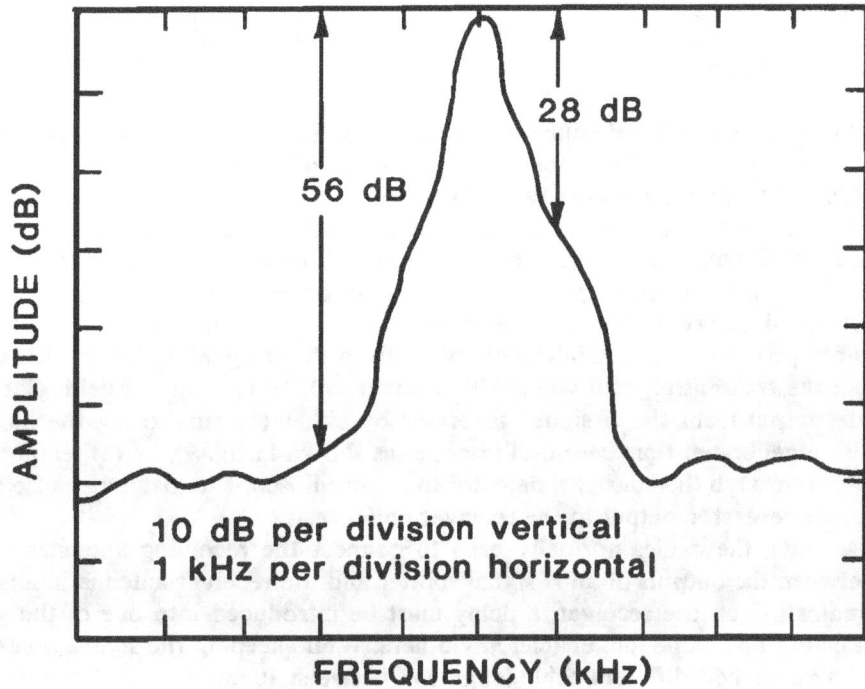

FIGURE 8. *Typical spectrum analyzer display obtained in measuring sideband spectrum reduction.*

5.7 Transmitter Antenna Radiation Efficiency Test (Type II Only)

Calculate the transmitter antenna radiation efficiency using the formula $E^2R^2/30P$, where E is the field strength of the carrier in volts per meter measured at a distance of R meters from the transmitter and P is the measured transmitter peak output power when terminated in a 50-Ω load [4].

The value for E was determined at a distance of 30 m from the transmitter during the measurement of radiated spurious emissions (see sec. 5.6.2.1). If the field strength is recorded in dBμV/m, the value of E can be calculated as

$$E = 10^{E'/20} \times 10^6$$

where E' is the field strength in dBμV/m. The value for P was determined in section 5.6.1.1.

5.8 Transmitter Battery Service Life Test

With the transmitter switched off, install fully-charged batteries of the type specified by the manufacturer, or equivalent, in the transmitter. Check battery charge by subjecting each battery to a 6-s maximum transmitter current drain.[3] If the output voltage of the charged battery decreases more than 5 percent during the momentary current drain, replace the battery with another fully-charged battery and repeat this procedure until a satisfactory result is obtained.

Connect the spectrum analyzer to the transmitter output and switch on the transmitter. Leave it on until the test is completed, measuring the elapsed time the transmitter is on. After 10 min, measure the transmitter peak output power by the method described in section 5.6.1.1. Then record the time required for the transmitter output power to decrease by 3 dB and then by 10 dB. The power measuring device need not be connected continuously during the test, *but if it is disconnected, a dummy 50-Ω load must be connected to the transmitter output* to prevent possible damage to the transmitter output circuits.

---

[3] If there is only one battery or the batteries are connected in series, each should be subjected to the maximum transmitter current drain. If the batteries are connected in parallel, each should be subjected to a current equal to the maximum divided by the number of batteries in parallel.

5.9 Receiver Tests

5.9.1 Sensitivity Tests

Connect the receiver and test equipment as shown in figure 9 for receivers using an antenna that does not include signal processing. If signal processing is a function of the receiving antenna, insert the antenna portion producing it into the signal path as shown. If a separate frequency counter is used, couple it to the rf signal generator output by means of an additional signal splitter or sampler in order to maintain a constant 50-Ω load impedance for the rf signal generator. A pulse generator is necessary to modulate the rf generator only if the rf generator itself is not capable of providing the proper pulse-modulated output.

Set the rf signal generator to one of the receiver standard test frequencies. Using the oscilloscope, adjust the equipment to produce a pulse-modulated output from the rf signal generator that approximates the pulsed rf output from the transmitter with which the receiver will be used in the field. Check the pulse width and repetition rate output from the rf signal generator by either (1) monitoring the modulating pulse into the generator with a calibrated time base oscilloscope (as shown in fig. 9), or (2) temporarily feeding the signal generator output through the coaxial rf detector to the oscilloscope to make the pulse measurements, and then reconnecting the generator output to the receiver under test.

For these tests, the cables normally used to connect the receiving antennas to the receiver shall be connected between the outputs of an rf signal splitter and the receiver antenna inputs. However, to produce a direction indication on the receiver, a delay must be introduced into one of the splitter-to-receiver connections by making that cable longer than any others. While keeping the signal generator output constant at some minimum value, add different cable lengths of between 10 and 45 cm (4 and 18 in) to one of the signal splitter-to-receiver connections, and note which added cable length produces the largest direction indication. Install this added cable length for the following tests. Switch any receiver controls that allow attenuation of the input signal to the nonattenuating position.

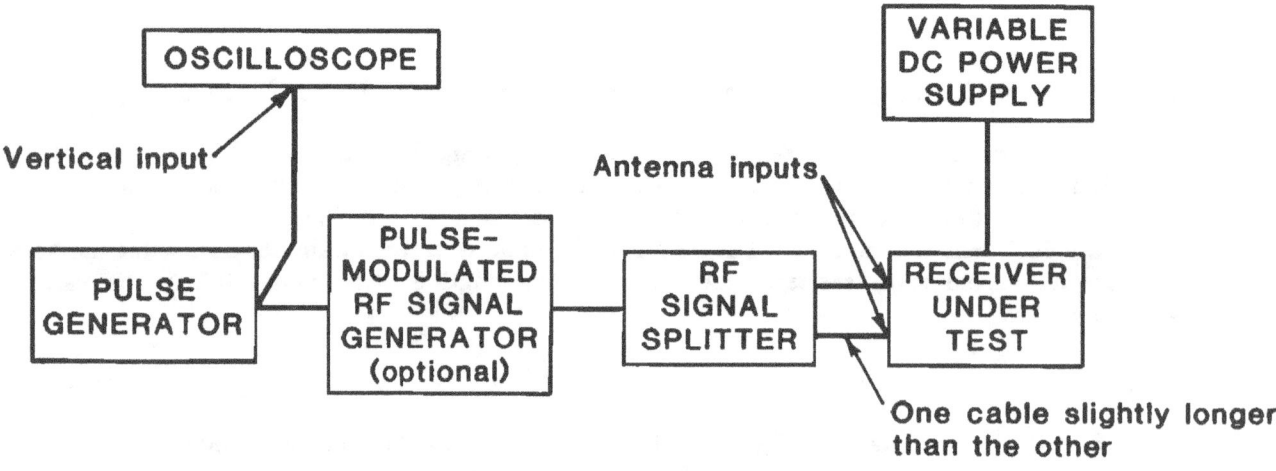

FIGURE 9. *Block diagram for direction indication sensitivity, distance indication sensitivity, and dynamic range measurements.*

5.9.1.1 Direction Indication Sensitivity Test

Apply standard supply voltage to the receiver and change the signal generator output amplitude until a direction indication equal to 10 percent of full scale or twice the amplitude of random noise variations, whichever is greater, has been achieved. Next, vary the receiver standard supply voltage from 11.4 to 14.4 V, stopping at that voltage which results in minimum deflection of the direction indication device. Readjust the signal generator output, if necessary, to achieve the specified deflection. Read the output from the signal generator in dBm. This value, after correction for signal-splitter insertion loss and any signal loss due to long cable lengths, is the receiver direction indication sensitivity.

Signal-splitter insertion loss depends upon the number of output ports. Theoretical loss is 3 dB between the input and outputs for two outputs, 4.8 dB for three outputs, 6 dB for four outputs, etc. Check the specifications for the unit being used and apply the appropriate factor. Cable lengths exceeding 20 ft can also add measurable loss. Determine cable loss by measuring signal amplitude, in decibels, at the generator output terminal and at the end of the cable used and calculate the difference.

5.9.1.2 Distance Indication Sensitivity Test

Apply standard supply voltage to the receiver and change the signal generator output amplitude until a distance indication equal to at least 10 percent of full scale has been achieved. Next, vary the receiver standard supply voltage from 11.4 to 14.4 V, stopping at that voltage which results in a minimum distance indication. Readjust the signal generator, if necessary, to achieve the specified distance indication. Read the output from the signal generator in dBm. This value, after correction for signal splitter insertion loss and any signal loss due to long cable lengths, as described in section 5.9.1.1, is the receiver distance indication sensitivity.

5.9.2 Selectivity Tests

Receiver selectivity characteristics are determined by measuring receiver sensitivity for various input frequencies and comparing the results.

Connect the receiver and test equipment as shown in figure 10, connecting the cables between the rf signal splitter and receiver under test as described in section 5.9.1.

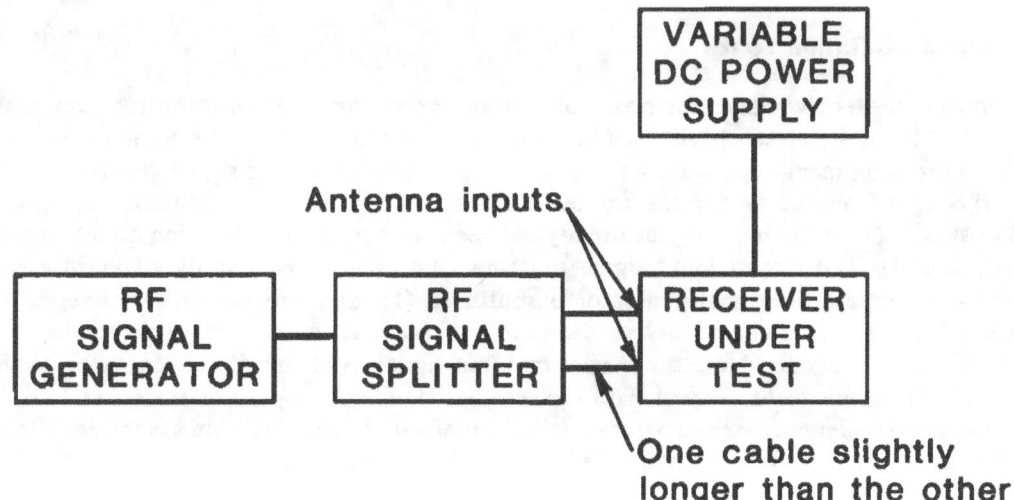

FIGURE 10. *Block diagram for adjacent-channel attenuation and spurious response attenuation measurements.*

5.9.2.1 Adjacent-Channel Attenuation Test

Set the signal generator frequency to one of the receiver standard test frequencies and apply standard supply voltage to the receiver. Vary the signal generator frequency slightly in each direction to find the point at which the direction indication is a maximum. Leave the frequency set at that point and increase the amplitude of the signal generator output until a direction indication of 10 percent of full scale (or some other convenient reference point, as long as it is near the lower end of the scale) is achieved. Record the output from the signal generator in dBm.

Next, increase the signal generator frequency by an amount equal to the specified channel spacing (see sec. 2) and adjust the signal generator output until the reference direction indication is again achieved. Record the output from the rf signal generator in dBm. Calculate the difference between the two readings. Then decrease the signal generator frequency so it is below the specified receiver frequency by an amount equal to the channel spacing and repeat the above procedure.

Set the rf signal generator to the offset frequency which resulted in the smallest difference in dBm, and adjust the amplitude of the signal generator output to again produce the reference point direction indication. Then vary the receiver standard supply voltage from 11.4 to 14.4 V dc and note whether the direction indication changes significantly. If it does, repeat the measurements just described at different supply voltages in order to find the smallest difference in dBm. This difference is the adjacent-channel attenuation.

5.9.2.2 Spurious Response Attenuation Test

Using the test setup shown in figure 10, set the signal generator frequency to one of the receiver standard test frequencies and apply standard supply voltage to the receiver. Then increase the signal generator output until a direction indication of 10 percent of full scale (or some other convenient reference point as long as it is near the lower end of the scale) is achieved. Record the output from the signal generator in dBm.

Decrease the signal generator frequency to a value below the lowest intermediate frequency of the receiver and increase the signal generator output amplitude by 70 dB. *Slowly* increase the signal generator frequency, stopping at any frequency that causes a change on the direction indication meter. At any such frequency, adjust the signal generator output to re-establish the reference deflection on the direction indication device and record the output from the signal generator in dBm. Repeat this process until the signal generator frequency is at least 1000 MHz.

Calculate the difference in dBm between each reading obtained with the signal generator set to the receiver standard test frequency and each reading obtained with it set to each of the other frequencies that caused an indication. Ignore any frequency that is a harmonic of the specified frequency. Reset the signal generator to the spurious frequency that produced the smallest difference and adjust its output to produce the reference point direction indication. Then vary the receiver standard supply voltage from 11.4 to 14.4 V dc and note whether the deflection changes significantly. If it does, locate the supply voltage at which the deflection is greatest without changing the signal generator frequency. Then repeat the measurements of signal generator output needed to establish the reference deflection for the specified receiver frequency and for the afore-mentioned spurious frequency. Calculate the difference between these two signal generator outputs in dBm. The smallest difference obtained is the spurious response attenuation.

5.9.3 Dynamic Range Test

Connect the receiver and test equipment as shown in figure 9, with the receiver sensitivity control, if available, set to minimum sensitivity. Set the rf signal generator to one of the receiver standard test frequencies and adjust the equipment to produce a pulse-modulated output from the signal generator that approximates the pulsed rf output from the system transmitter. Refer to section 5.9.1 for additional set-up information.

The smallest receiver input signal already has been measured in determining direction indication sensitivity (see sec. 5.9.1.1). The largest input signal must now be determined. Apply standard supply voltage to the receiver and increase the signal generator output until (1) there are no further changes in relative distance indications[4], (2) the change in relative distance indication is in the wrong direction, or (3) the direction indication becomes erratic. Note the maximum input signal from the signal generator in dBm.

Repeat the above measurement with the receiver standard supply voltage set at 11.4, 12.5, 13.8 and 14.4 ✦ V dc. Record the *lowest* maximum input signal obtained. From this value, subtract the receiver direction indication sensitivity in dBm to obtain the dynamic range in decibels.

5.10 Transmitter/Receiver Figure of Merit Test

Add the transmitter peak output power in dBm measured with the transmitter supply voltage set at 20 percent below the nominal value (sec. 5.6.1.1) to the absolute value of the minimum receiver direction indication sensitivity in dBm (sec. 5.9.1.1) to obtain the transmitter/receiver figure of merit.

---

[4] On some receivers, the relative distance is indicated in discrete steps, e.g. , by three or four different lights, each indicating a different distance range. Under these conditions, the input signal which activates the closest range indicator is the maximum input signal as far as the input voltage range calculation is concerned.

APPENDIX A—REFERENCES

[1]   Greene, F. M. Technical terms and definitions used with law enforcement communications equipment. LESP-RPT-0203.00. National Institute of Justice, U.S. Department of Justice, Washington, DC 20531; 1973 June.

[2]   Private land mobile radio services, section 90.19, police radio service. Rules and Regulations, Vol. 5, Part 90. Federal Communications Commission, 1919 M Street, NW., Washington, DC 20554.

[3]   Frequency allocations and radio treaty matters; general rules and regulations. Rules and Regulations, Vol. 5, Part 2. Federal Communications Commission, 1919 M Street, NW., Washington, DC 20554.

[4]   Reference data for radio engineers. 6th ed. New York: Howard W. Sams and Co. Inc.; 1975: Chapter 27, Antennas, p. 27–7.

APPENDIX B—BIBLIOGRAPHY

[1] American national standard specifications for dry cells and batteries. ANSI Standard C18.1-1979. American National Standards Institute; 1979 May.

[2] Batteries for personal/portable transceivers. NILECJ-STD-0211.00. National Institute of Justice, U.S. Department of Justice, Washington, DC 20531; 1975 June.

[3] Batteries used with law enforcement communications equipment: comparison and performance characteristics. LESP-RPT-0201.00. National Institute of Justice, U.S. Department of Justice, Washington, DC 20531; 1972 May.

[4] Body-worn FM transmitters. NILECJ-STD-0214.00. National Institute of Justice, U.S. Department of Justice, Washington, DC 20531; 1978 December.

[5] Eveready battery engineering data. Union Carbide Corp., New York, NY; 1976.

[6] IEEE standard dictionary of electrical and electronic terms. IEEE Standard 100-1977. Institute of Electrical and Electronic Engineers, Inc., 345 East 47th St., New York, NY; 1977.

[7] IEEE standard report on measuring field strength in radio wave propagation. IEEE Standard 291-1969. Institute of Electrical and Electronic Engineers, Inc., 345 East 47th St., New York, NY; 1969 May.

[8] IEEE standard test procedures for antennas. IEEE Standard 149–1979. Institute of Electrical and Electronic Engineers, Inc., 345 East 47th St., New York, NY; 1979 December.

[9] Lapedes, Daniel N., ed. McGraw-Hill dictionary of scientific and technical terms. 2nd ed. New York: McGraw-Hill Book Co.; 1978.

[10] Minimum standards for land-mobile communication antennas, Part II—Vehicular antennas. EIA Standard RS-329-1. Electronic Industries Association, 2001 Eye St., NW., Washington, DC 20006; 1972 August.

[11] Minimum standards for land-mobile communication FM or PM receivers, 25–947 MHz. EIA Standard RS-204-C. Electronic Industries Association, 2001 Eye St., NW., Washington, DC 20006; 1982 January.

[12] Minimum standards for land-mobile communication FM or PM transmitters, 25–470 MHz. EIA Standard RS-152-B. Electronic Industries Association, 2001 Eye St., NW., Washington, DC 20006; 1970 February.

[13] Minimum standards for portable/personal radio transmitters, receivers, and transmitter/receiver combination land-mobile communication FM or PM equipment, 25–1000 MHz. EIA Standard RS-316-B. Electronic Industries Association, 2001 Eye St., NW., Washington, DC 20006; 1979 May.

[14] Mobile antennas. NILECJ-STD-0205.00. National Institute of Justice, U.S. Department of Justice, Washington, DC 20531; 1974 May.

[15] Personal FM transceivers. NILECJ-STD-0209.01. National Institute of Justice, U.S. Department of Justice, Washington, DC 20531; (In Press).

[16] Spectrum analysis-field strength measurement. Hewlett Packard Application Note 150-10; 1976 September.

[17] Spectrum analysis... spectrum analyzer basics. Hewlett-Packard Application Note 150; 1974 April.

[18] Straus, Isidor. Testing products correctly ensures EMI-spec compliance. EDN. 1981 November 25: 121–130.